MANOVA by Example
Hands on Approach using SAS

Faye Anderson, MS, PhD

Copyright © 2019 Author Name

All rights reserved.

ISBN: 9781674891378

DEDICATION

To my family.

Table of Contents

Chapter 1: Overview of Hypothesis Testing	5
p-value Interpretation	5
Commonly used Hypothesis Tests	6
T-test in SAS	7
Chapter 2: ANOVA vs ANCOVA vs MANOVA vs MANCOVA	10
What is ANOVA?	10
ANOVA Assumptions	11
One-way ANOVA Example	11
Air Quality Dataset	11
Two-way ANOVA Example	16
Membership Dataset	16
ANCOVA and its Assumptions	19
ANCOVA Example	19
Summary	21
Chapter 3: One-way MANOVA	23
What is MANOVA?	23
MANOVA Assumptions	23
One-way MANOVA	24
Two-way MANOVA	27
One-way vs Two-way MANOVA	30
Chapter 4: MANOVA and Interaction	32
What is Interaction?	32
Association versus Interaction	34
MANOVA and Interaction	34
ABOUT THE AUTHOR	38

Chapter 1: Overview of Hypothesis Testing

Since analysis of variance is a type of hypothesis testing, it is imperative to have an overview of hypothesis testing.

In studies where we are comparing groups, we statistically test the difference between them by conducting hypothesis testing. An example would be the difference between groups of patients: one took a placebo, another took drug A, and another took drug B.

In hypothesis testing, there are two hypotheses: the null hypothesis (H0) and the alternative hypothesis (Ha). In this context, H0 states that the population parameter (e.g. the mean or variance) is equal for the groups. Whereas the null hypothesis states that the groups' parameters are not equal.

SAS offers free license at the following link:

https://www.sas.com/en_us/software/university-edition/download-software.html

p-value Interpretation

Every hypothesis test produces a probability value or p-value which assesses how compatible the sample data are with the null hypothesis. High p-values indicate that the data is compatible with H0, whereas low p-values indicate that the data is not compatible with H0 (or agrees with Ha).

In order to understand how to interpret p-value, it is important to get acquainted with the concept of significance level (alpha or critical value). This is the probability of rejecting the null hypothesis when it is true, or the probability of committing a type I error. With a significance level of 0.05, there is a 5% chance of rejecting a true null hypothesis. If the significance level is higher than the conventional 0.05, such as 0.10, this will increase the chance of a false positive to 0.10, but it will also decrease the chance of a false. If the significance level is lower than the conventional 0.05, such as 0.01, this decreases the chance of a false positive.

The following table demonstrates the four scenarios between having true (or not true) H0 and making the correct or incorrect decision. Alpha is the probability of committing type I error or rejecting a true H0 whereas beta is the probability of committing type II error or failing to reject a false H0. The statistical power of the test is equal to 1 - beta.

	H0 True	H0 False
Decision 1: Fail to reject H0	Correct Decision	Incorrect Decision = Type II Error (Beta)
Decision 2: Reject H0	Incorrect Decision = Type I Error (Alpha)	Correct Decision

The following table summarizes how to interpret p-value in general.

Scenario	Decision	Conclusion
p-value <= significance level (0.05)	Reject the null hypothesis (H0).	Ha is true
p-value > significance level (0.05)	Fail to reject the null hypothesis (H0).	H0 is true.

Commonly used Hypothesis Tests

This section demonstrates the following important four tests.

Test	Purpose
T-test	Compare the means of two populations
F-test	Compare the variances of two populations
Kolmogorov-Smirnov	Tests whether the data is normally distributed or not, when the sample size is large or when there are no outliers in the sample.
Shapiro-Wilk	Tests whether the data is normally distributed or not, when the sample size is relatively small (n < 50) or when there are outliers in the sample.
Levene's	Tests whether the groups' variances are equal or not

T-test in SAS

The following code demonstrates how to use proc ttest in SAS to conduct t-test. The output shows the p-value of <.0001, which is less than the default significance level of 0.05. So, we reject the null hypothesis H0. In this example, H0 states that the mean of the variable year-rev is equal to zero. We conclude that the mean of year_rev is not 0.

One-Sample t-test Example:

```
/* t test */
proc ttest data=WORK.IMPORT1 sides=2 h0=0 plots=none;
        var year_rev; run;
```

Output:

Variable: year_rev

N	Mean	Std Dev	Std Err	Minimum	Maximum
51	65.8765	6.7344	0.9430	37.4000	75.9000

Mean	95% CL Mean		Std Dev	95% CL Std Dev	
65.8765	63.9824	67.7706	6.7344	5.6347	8.3714

DF	t Value	Pr > |t|
50	69.86	<.0001

The following example demonstrates how proc ttest can be used to compare the means of two variables/groups ozone and wind. T-test assumes the variances of the two groups are equal. Since p-value is less than 0.05, we reject the null hypothesis that the two means for ozone and wind are equal.

Proc univariate was used to test for normality. The small p-values (less than 0.05) force us to reject the null hypothesis of normality for the difference between ozone and wind.

Two-Sample t-test Example:

```
data Work._Paired_diffs_;
        set BOOK.AIRQUALITY;
```

```
            _Difference_=Ozone - Wind;
            label _Difference_="Difference: Ozone - Wind";
run;

/* Test for normality */
proc univariate data=Work._Paired_diffs_ normal mu0=0;
            ods select TestsForNormality;
            var _Difference_;
run;
```

Output:

Variable: _Difference_ (Difference: Ozone - Wind)

Tests for Normality				
Test		Statistic		p Value
Shapiro-Wilk	W	0.889226	Pr < W	<0.0001
Kolmogorov-Smirnov	D	0.153452	Pr > D	<0.0100
Cramer-von Mises	W-Sq	1.061138	Pr > W-Sq	<0.0050
Anderson-Darling	A-Sq	5.810486	Pr > A-Sq	<0.0050

Difference: Ozone - Wind

N	Mean	Std Dev	Std Err	Minimum	Maximum
153	29.2320	31.0456	2.5099	-12.4000	130.9

Mean	95% CL Mean		Std Dev	95% CL Std Dev	
29.2320	24.2733	34.1908	31.0456	27.9134	34.9758

DF	t Value	Pr > |t|
152	11.65	<.0001

Chapter 2: ANOVA vs ANCOVA vs MANOVA vs MANCOVA

What is ANOVA?

Analysis of variance also known as ANOVA is a type of statistical hypothesis testing that compares two or more groups' means. There are three types of ANOVA: one-way ANOVA, two-way ANOVA, and k-way ANOVA. Some books classify ANOVA in two categories by combining the two later types into one group. The table below demonstrates the differences between these types. The following table is from my book on ANOVA using SAS.

ANOVA Type	Uses	Example
One-way ANOVA	When we need to compare only one factor or independent variable between the groups.	If we want to compare whether the mean pay of three workers is the same based on their working hours.
Two-way ANOVA	When we need to compare two factors between the groups.	If we want to compare whether the mean pay of three workers is the same based on two factors: their working hours and years of education.
K-way ANOVA	When we need to compare k (more than 2) factors between the groups.	If we want to compare whether the mean pay of three workers is the same based on few factors: their working hours, gender, number of previous jobs, and years of education.

ANOVA Assumptions

ANOVA is a parametric test. This means that its assumptions need to be verified prior to running the analysis. The three assumptions of ANOVA are: observations are independent (data was sampled randomly), data is normally distributed, and homogeneity (the equality of the population variances for the groups).

The result of an ANOVA analysis is that there is a significant difference between groups, not which groups are significantly different from each other. This limitation can be alleviated by conducting a post-hoc comparison to find out where the differences are significantly different from each other and which are not. Example post-hoc comparisons are Scheffe's and Tukey's.

One-way ANOVA Example
Air Quality Dataset
This dataset contains daily air quality measurements in New York, May to September 1973. It can be downloaded from R:

https://www.rdocumentation.org/packages/datasets/versions/3.6.1/topics/airquality

The dataset has four continuous variables for the measurements of ozone, solar radiation, wind, and temperature. It has two categorical variables: month and day. The data was collected for 5 months, from May through September. Following are the summary statistics.

Analysis Variable : Ozone Ozone				
Mean	Std Dev	N	Skewness	Kurtosis
39.1895425	29.3333275	153	1.1754125	0.8058269

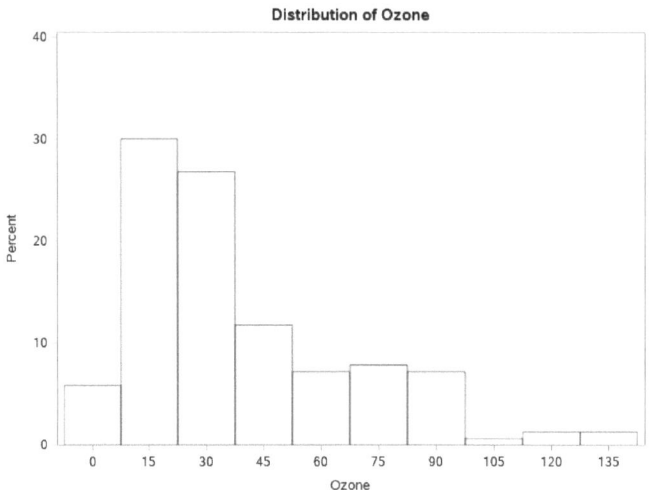

Before we conduct one-way ANOVA to test if ozone levels were equal in the five months, we need to verify the three assumptions of randomness, independence, and homogeneity.

Randomness is assumed in this documented study. It is obvious that ozone is not normally distributed because of the skewness of its histogram. Moreover, its skewness is far from zero and its kurtosis is far from 3. Thus, we transform it using the natural logarithm function before conducting three types of ANOVA: using PROC GLM, using PROC ANOVA, and using PROC npar1way.

The null hypothesis for this procedure states that the monthly means for ozone are equal from May through September. The alternative hypothesis states that at least one month's mean is different from the remaining four.

Transforming data:

```
data work.transform;
        set WORK.AIRQUALITY;
        log_Ozone=log(Ozone);
run;
```

One-way ANOVA Example using Proc GLM:

```
ods noproctitle;
ods graphics / imagemap=on;

proc glm data=WORK.TRANSFORM plots=none;
        class Month;
        model log_Ozone=Month;
        means Month / hovtest=levene plots=none;
        lsmeans Month / plots=none;
        run;
quit;
```

Output:

Class Level Information		
Class	Levels	Values
Month	5	5 6 7 8 9

Number of Observations Read	153
Number of Observations Used	153

Dependent Variable: log_Ozone

Source	DF	Sum of Squares	Mean Square	F Value	Pr > F
Model	4	18.0280650	4.5070163	7.94	<.0001
Error	148	84.0373784	0.5678201		
Corrected Total	152	102.0654434			

R-Square	Coeff Var	Root MSE	log_Ozone Mean
0.176632	22.32070	0.753538	3.375962

Source	DF	Type I SS	Mean Square	F Value	Pr > F
Month	4	18.02806501	4.50701625	7.94	<.0001

Source	DF	Type III SS	Mean Square	F Value	Pr > F
Month	4	18.02806501	4.50701625	7.94	<.0001

Levene's Test for Homogeneity of log_Ozone Variance ANOVA of Squared Deviations from Group Means					
Source	DF	Sum of Squares	Mean Square	F Value	Pr > F
Month	4	2.2654	0.5663	0.70	0.5921
Error	148	119.5	0.8072		

Level of Month	N	log_Ozone	
		Mean	Std Dev
5	31	2.80184426	0.89379840
6	30	3.44318004	0.76204484
7	31	3.79463204	0.70023652
8	31	3.60494963	0.72770282
9	30	3.23275442	0.66004115

Least Squares Means

Month	log_Ozone LSMEAN
5	2.80184426
6	3.44318004
7	3.79463204
8	3.60494963
9	3.23275442

One-way ANOVA Example using Proc ANOVA:

PROC ANOVA DATA=WORK.TRANSFORM;
CLASS Month;
MODEL log_Ozone = Month;
TITLE 'Compare Ozone across Months';
RUN;;

Output:

Compare Ozone across Months

Class Level Information

Class	Levels	Values
Month	5	5 6 7 8 9

Number of Observations Read	153
Number of Observations Used	153

Compare Ozone across Months

Dependent Variable: log_Ozone

Source	DF	Sum of Squares	Mean Square	F Value	Pr > F
Model	4	18.0280650	4.5070163	7.94	<.0001
Error	148	84.0373784	0.5678201		
Corrected Total	152	102.0654434			

R-Square	Coeff Var	Root MSE	log_Ozone Mean
0.176632	22.32070	0.753538	3.375962

Source	DF	Anova SS	Mean Square	F Value	Pr > F
Month	4	18.02806501	4.50701625	7.94	<.0001

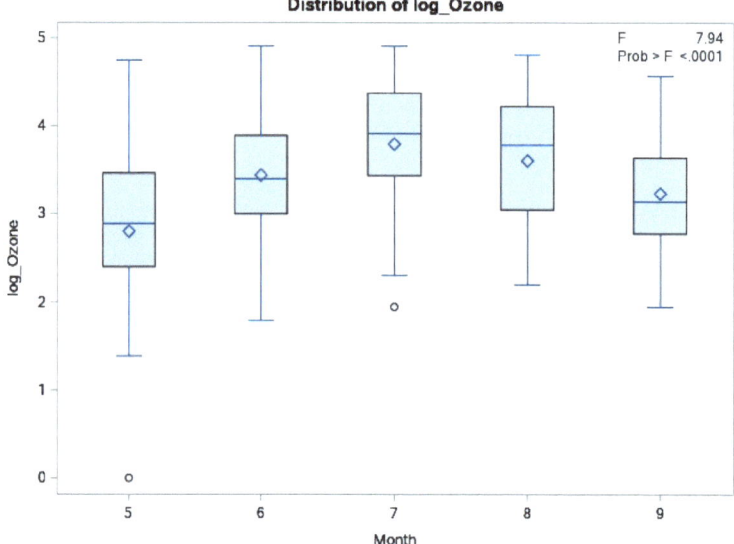

Two-way ANOVA Example

In two-way ANOVA, there are two categorical independent variables that we suspect they impact the independent variable or the outcome.

Membership Dataset

The dataset Membership has two continuous variables duration and price; and two categorical variables plan and area. The table below presents the description of the variables.

Variable	Description
Membership	Number of weeks the user has been a

		member with the company
Area		Categorical variable for the three areas of service: A, B, and C.
Price		Amount of money the customer pays weekly.
Plan		Categorical variable for whether the customer is subscribed with the plan: 1 for yes and 0 for no.

In the Membership dataset, we would like to investigate the impact of both the area and plan on duration of membership. Here we have three null hypotheses:

First H0: The means of Duration in weeks grouped by plan are the same. There are 2 means for Duration based on the two options for Plan.

Second H0: The means of Duration in weeks grouped by Area are the same. There are 3 means for Duration based on the three areas A, B, and C.

Third H0: There is no interaction between Area and Plan to impact the mean Duration in Weeks.

```
ods noproctitle;
ods graphics / imagemap=on;

proc anova data=WORK.MEMBERSHIP;
    class Plan Area;
        model Duration = Plan Area Plan*Area;
```

run;

Output:

Class Level Information		
Class	Levels	Values
Plan	2	0 1
Area	3	A B C
Number of Observations Read	3333	
Number of Observations Used	2151	

Dependent Variable: Duration Duration

Source	DF	Sum of Squares	Mean Square	F Value	Pr > F
Model	5	5505.742	1101.148	0.68	0.6371
Error	2145	3463350.166	1614.615		
Corrected Total	2150	3468855.908			

R-Square	Coeff Var	Root MSE	Duration Mean
0.001587	39.87694	40.18228	100.7657

Source	DF	Anova SS	Mean Square	F Value	Pr > F
Plan	1	48.273626	48.273626	0.03	0.8627
Area	2	5352.331625	2676.165812	1.66	0.1909
Plan*Area	2	105.136320	52.568160	0.03	0.9680

Interpretation of Results:

The "model" statement lists Duration as dependent variable, Plan and Area as independent variable. Furthermore, Plan*Area represents the interactions between Plan and Area.

The last table of the results gives the p-values which are all greater than alpha of 0.05. That is, we fail to reject the three null hypotheses that the three factors of Plan, Area, and Plan*Area do not have a statistically significant effect on the duration of membership (weeks). In other words, the three null hypotheses are true.

ANCOVA and its Assumptions

ANCOVA is "ANOVA with covariates" or, more simply, a mix of ANOVA and linear regression. In analysis of covariance (ANCOVA), a continuous variable (the covariate) is introduced into the model of an analysis-of-variance experiment.

In addition to the three assumptions of ANOVA, ANCOVA requires the following three assumptions:

1. The relationship between the dependent variable (y) and the covariate (x) is linear.
2. The lines expressing these linear relationships are all parallel.
3. The covariant and independent variables are independent.

ANCOVA Example

For the airquality dataset, the data was collected from May through September of the same year. We would like to explore the relationship between month and temperature on ozone using ANCOVA. Here the dependent variable is ozone, the independent variable is month, and the covariate is temperature.

```
title;
ods noproctitle;
ods graphics / imagemap=on;

proc glm data=AIRQUALITY;
```

```
        class Month;
        model Ozone=Month Temp Temp * Month;
quit;
```

Class Level Information		
Class	Levels	Values
Month	5	5 6 7 8 9
Number of Observations Read		153
Number of Observations Used		153

Dependent Variable: Ozone Ozone

Source	DF	Sum of Squares	Mean Square	F Value	Pr > F
Model	9	47842.5662	5315.8407	9.16	<.0001
Error	143	82944.9371	580.0345		
Corrected Total	152	130787.5033			

R-Square	Coeff Var	Root MSE	Ozone Mean
0.365804	61.45493	24.08391	39.18954

Source	DF	Type I SS	Mean Square	F Value	Pr > F
Month	4	18429.95058	4607.48764	7.94	<.0001
Temp	1	20311.20939	20311.20939	35.02	<.0001
Temp*Month	4	9101.40624	2275.35156	3.92	0.0047

Source	DF	Type III SS	Mean Square	F Value	Pr > F
Month	4	9133.80341	2283.45085	3.94	0.0046
Temp	1	22285.94963	22285.94963	38.42	<.0001
Temp*Month	4	9101.40624	2275.35156	3.92	0.0047

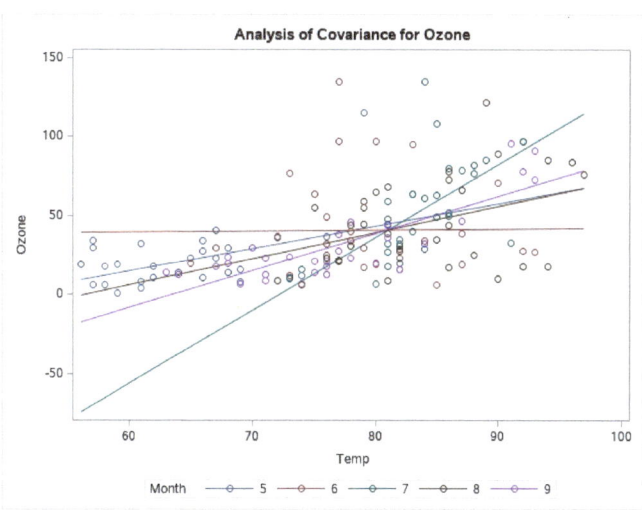

Interpretation of the Results:

In this example we compare the levels of ozone under the influence of temperature over the five months. The null hypothesis states that there is no significant effect of temperature on ozone controlling for month. This code investigated the Ozone=Month +Temp + Temp * Month. Where temp*Month is the interaction between month and temperature. The p-values for the three factors of month, temperature, and month*temperature were less than 0.05. Hence, they are all statistically significant to ozone levels. The p-value for the model was < 0.0001, which means that the model is statistically significant.

Summary

The following table summarizes the main differences between ANOVA, ANCOVA, MANOVA, and MANCOVA. The latter two will be demonstrated in the following chapters.

ANCOVA and MANCOVA have continuous independent variables or covariates. ANOVA and MANOVA have categorical independent variables.

The letter M for multivariate refers to having more than one dependent variable (MANOVA and MANCOVA).

	Continuous Independent Variable (covariates)	Categorical Independent Variable	Number of dependent variables	Number of independent variables
ANOVA		X	1	1
ANCOVA	X		1	2
MANOVA		X	2 or more	1
MANCOVA	X		2 or more	Multiple covariates

Chapter 3: One-way MANOVA

Before we learn about the different types of MANOVA, let us define it and learn about its assumptions.

What is MANOVA?

MANOVA is an ANOVA with multiple dependent variables. ANOVA tests for the difference in means between two or more groups, whereas MANOVA tests for the difference in two or more vectors of means. The purpose of MANOVA is to establish if the response variables (e.g. ozone and carbon dioxide levels in the air) are altered by manipulation of the independent variable temperature.

MANOVA Assumptions

In addition to the assumptions of ANOVA, MANOVA has the following four assumptions, along with suitable tests:

Assumption	Test
Absence of multivariate outliers	Mahalanobis Distances
Linearity	Scatterplot matrix
Absence of multicollinearity	Correlation matrix
Equality of covariance metrices	Box's M test

There are two main types of MANOVA: one-way MANOVA and two-way MANOVA.

One-way MANOVA

In the following example, using the airquality dataset, we are testing the hypothesis that the independent variable Month impacts the two dependent variables of ozone and temperature.

Title;
ods noproctitle;
ods graphics / imagemap=on;

proc glm data=AIRQUALITY;
    class Month;
    model Ozone Temp = Month;
    manova h =_all_ ;

    run;
quit;

Class Level Information

Class	Levels	Values
Month	5	5 6 7 8 9

Number of Observations Read	153
Number of Observations Used	153

Dependent Variable: Ozone Ozone

Source	DF	Sum of Squares	Mean Square	F Value	Pr > F
Model	4	18429.9506	4607.4876	6.07	0.0002
Error	148	112357.5527	759.1727		
Corrected Total	152	130787.5033			

R-Square	Coeff Var	Root MSE	Ozone Mean
0.140915	70.30725	27.55309	39.18954

Source	DF	Type I SS	Mean Square	F Value	Pr > F
Month	4	18429.95058	4607.48764	6.07	0.0002

Source	DF	Type III SS	Mean Square	F Value	Pr > F
Month	4	18429.95058	4607.48764	6.07	0.0002

MANOVA by Example

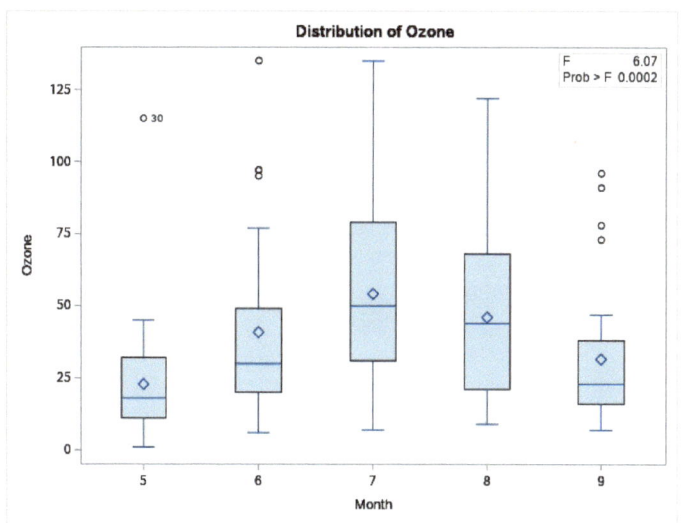

Dependent Variable: Temp Temp

Source	DF	Sum of Squares	Mean Square	F Value	Pr > F
Model	4	7061.12751	1765.28188	39.85	<.0001
Error	148	6556.75484	44.30240		
Corrected Total	152	13617.88235			

R-Square	Coeff Var	Root MSE	Temp Mean
0.518519	8.546230	6.656005	77.88235

Source	DF	Type I SS	Mean Square	F Value	Pr > F
Month	4	7061.127514	1765.281879	39.85	<.0001

Source	DF	Type III SS	Mean Square	F Value	Pr > F
Month	4	7061.127514	1765.281879	39.85	<.0001

MANOVA by Example

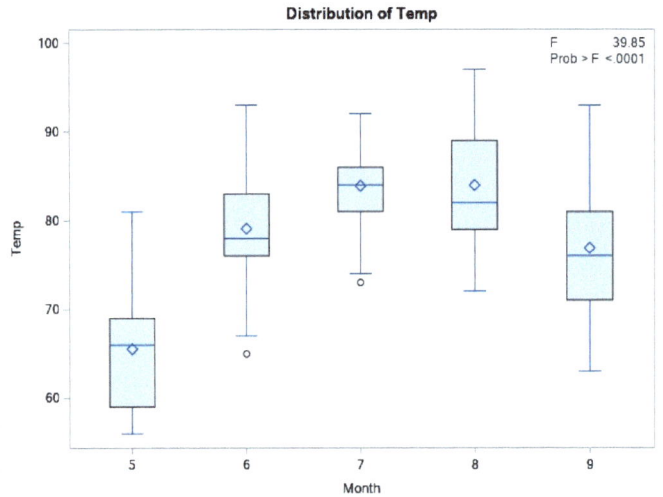

Multivariate Analysis of Variance

Characteristic Roots and Vectors of: E Inverse * H, where H = Type III SSCP Matrix for Month E = Error SSCP Matrix

		Characteristic Vector V'EV=1	
Characteristic Root	**Percent**	**Ozone**	**Temp**
1.08219076	97.55	-0.00023288	0.01272869
0.02717684	2.45	0.00328784	-0.00491418

MANOVA Test Criteria and F Approximations for the Hypothesis of No Overall Month Effect H = Type III SSCP Matrix for Month E = Error SSCP Matrix

S=2 M=0.5 N=72.5

Statistic	Value	F Value	Num DF	Den DF	Pr > F
NOTE: F Statistic for Roy's Greatest Root is an upper bound.					
NOTE: F Statistic for Wilks' Lambda is exact.					
Wilks' Lambda	0.46755668	17.00	8	294	<.0001
Pillai's Trace	0.54619440	13.90	8	296	<.0001
Hotelling-Lawley Trace	1.10936759	20.30	8	207.69	<.0001
Roy's Greatest Root	1.08219076	40.04	4	148	<.0001

Interpretation:
The last table shows significant p-values, which indicate that that differences between ozone and temperature depending on the month exist.

Two-way MANOVA

Two-way MANOVA is also called factorial MANOVA. It examines the joint effect of the independent variables on the dependent variables. It has three null hypotheses:

1. The means for the outcome (dependent variables) are the same across the categories of the first independent variable.
2. The means for the outcome (dependent variables) are the same across the categories of the second independent variable.
3. The interaction between the two independent variables is zero.

For the airquality data, we would like to investigate the impact of wind and month on ozone and temperature. The null hypotheses are as follows:

1. The means for ozone and temperature are the same across the five months.
2. The means for ozone and temperature are the regardless of the wind.
3. The interaction between the wind and month is zero.

Here is the code:

```
ods noproctitle;
ods graphics / imagemap=on;
proc glm data=MATH601.AIRQUALITY;
    class Wind Month;
    model Ozone Temp=Wind Month / ss1 ss3;
    model h = _all_;
quit;
```

Class Level Information

Class	Levels	Values
Wind	31	4 8 12 1.7 2.3 2.8 3.4 4.1 4.6 5.1 5.7 6.3 6.9 7.4 8.6 9.2 9.7 10.3 10.9 11.5 12.6 13.2 13.8 14.3 14.9 15.5 16.1 16.6 18.4 20.1 20.7
Month	5	5 6 7 8 9

Number of Observations Read	153
Number of Observations Used	153

Dependent Variable: Ozone Ozone

Source	DF	Sum of Squares	Mean Square	F Value	Pr > F
Model	34	63241.3199	1860.0388	3.25	<.0001
Error	118	67546.1833	572.4253		
Corrected Total	152	130787.5033			

R-Square	Coeff Var	Root MSE	Ozone Mean
0.483543	61.05050	23.92541	39.18954

Source	DF	Type I SS	Mean Square	F Value	Pr > F
Wind	30	54912.47953	1830.41598	3.20	<.0001
Month	4	8328.84040	2082.21010	3.64	0.0078

Source	DF	Type III SS	Mean Square	F Value	Pr > F
Wind	30	44811.36936	1493.71231	2.61	0.0001
Month	4	8328.84040	2082.21010	3.64	0.0078

MANOVA by Example

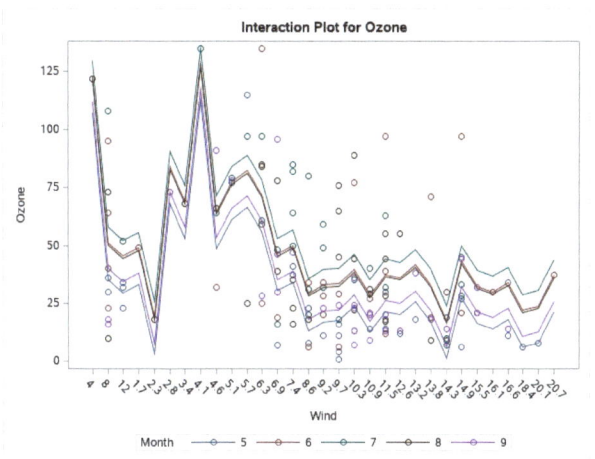

Dependent Variable: Temp Temp

Source	DF	Sum of Squares	Mean Square	F Value	Pr > F
Model	34	9034.99106	265.73503	6.84	<.0001
Error	118	4582.89130	38.83806		
Corrected Total	152	13617.88235			

R-Square	Coeff Var	Root MSE	Temp Mean
0.663465	8.001837	6.232019	77.88235

Source	DF	Type I SS	Mean Square	F Value	Pr > F
Wind	30	4456.845737	148.561525	3.83	<.0001
Month	4	4578.145321	1144.536330	29.47	<.0001

Source	DF	Type III SS	Mean Square	F Value	Pr > F
Wind	30	1973.863543	65.795451	1.69	0.0247
Month	4	4578.145321	1144.536330	29.47	<.0001

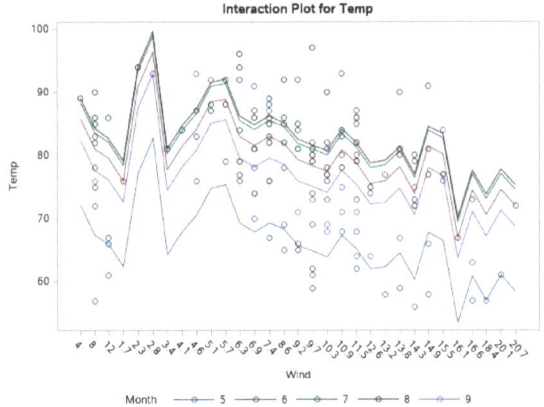

The p-value for the model (F test) is <0.0001, which is less than the significance level of 0.05. This indicates that this model is statistically significant. The p-values for wind and month are also less than 0.05, which indicates that they are statistically significant to the levels of ozone and temperature.

One-way vs Two-way MANOVA

Both methods have the same assumptions but slightly different application. The table below summarized the main difference between the two types: the number of independent variables or factors.

	Number of independent variables	Number of dependent variables	**Example**
One-way MANOVA	1	2+	Impact of temperature on air levels of ozone and carbon dioxide
Two-way MANOVA	2+	2+	Impact of both wind and temperature on air levels of ozone and carbon dioxide

Chapter 4: MANOVA and Interaction

Interaction arises when considering three or more variables. Hence, it is relevant in the context of a multivariate analysis such as MANOVA.

What is Interaction?

The interacting variable impacts how the independent variable affects the outcome or the dependent variable. For example, if a drug influences the body, and alcohol alone has another effect; then taking the drug along with alcohol will have a third effect that is not necessarily the addition of the two.

The table below presents the eight possibilities for pair-wise plots in terms of detecting interaction. The first four plots (parallel lines) show no interaction whereas the remaining four do.

MANOVA by Example

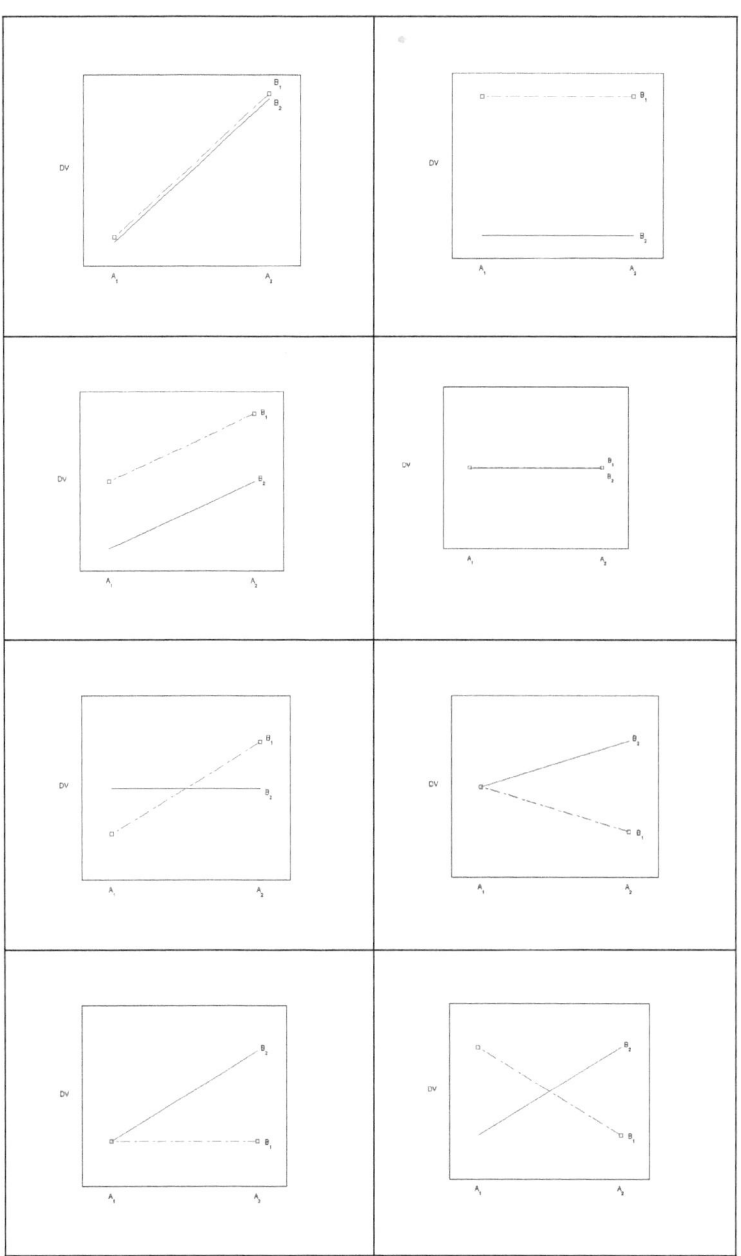

Association versus Interaction

Association between two variables means the values of one variable relate in some way to the values of the other. Association is usually measured by correlation for two variables (pair-wise). Association means the values of one variable generally co-occur with certain values of the other. Interaction is different than association. Whether two variables are associated says nothing about whether they interact in their effect on a third variable. Likewise, if two variables interact, they may or may not be associated. An interaction between two variables means the effect of one of those variables on a third variable is not constant—the effect differs at different values of the other. Best plots for interaction detection include scatter plot matrix, violin plots, and profile plot. Best plots for association detection include scatter plot matrix.

MANOVA and Interaction

One of the main uses of MANOVA is to investigate the presence of interaction between the independent variables on the dependent variables. To do this in SAS, the interaction term needs to be added to the model statement, in the form of multiplication/product of the terms. Here is an example.

Suppose that for the airquality data, we would like to investigate whether independent categorical variables Month and Day as well as their interaction impact the two dependent variables of ozone and temperature. The following is the code to do so.

```
Title;
ods noproctitle;
ods graphics / imagemap=on;

proc glm data=MATH601.AIRQUALITY;
    class Month Wind;
    model Ozone Temp =Month Wind Month*Wind;
    model h = _all_;
    run;
quit;
```

Class Level Information		
Class	Levels	Values
Month	5	5 6 7 8 9
Wind	31	4 8 12 1.7 2.3 2.8 3.4 4.1 4.6 5.1 5.7 6.3 6.9 7.4 8.6 9.2 9.7 10.3 10.9 11.5 12.6 13.2 13.8 14.3 14.9 15.5 16.1 16.6 18.4 20.1 20.7

Number of Observations Read	153
Number of Observations Used	153

Dependent Variable: Ozone Ozone

Source	DF	Sum of Squares	Mean Square	F Value	Pr > F
Model	87	93668.1366	1076.6452	1.89	0.0040
Error	65	37119.3667	571.0672		
Corrected Total	152	130787.5033			

R-Square	Coeff Var	Root MSE	Ozone Mean
0.716186	60.97803	23.89701	39.18954

Source	DF	Type I SS	Mean Square	F Value	Pr > F
Month	4	18429.95058	4607.48764	8.07	<.0001
Wind	30	44811.36936	1493.71231	2.62	0.0006
Month*Wind	53	30426.81667	574.09088	1.01	0.4884

MANOVA by Example

Source	DF	Type III SS	Mean Square	F Value	Pr > F
Month	4	4974.68580	1243.67145	2.18	0.0812
Wind	30	43157.29358	1438.57645	2.52	0.0010
Month*Wind	53	30426.81667	574.09088	1.01	0.4884

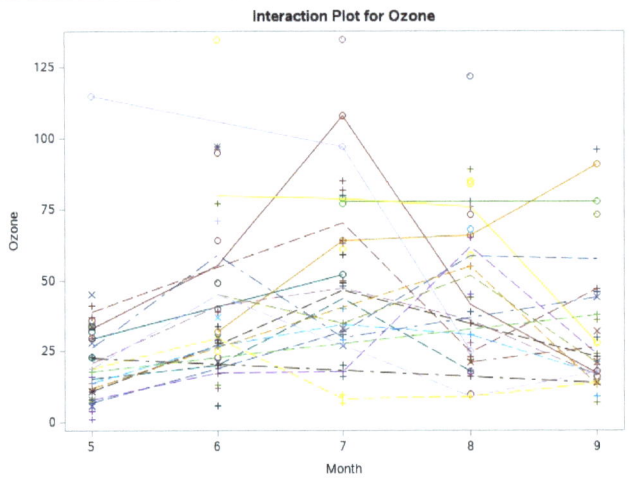

Dependent Variable: Temp Temp

Source	DF	Sum of Squares	Mean Square	F Value	Pr > F
Model	87	11117.09902	127.78275	3.32	<.0001
Error	65	2500.78333	38.47359		
Corrected Total	152	13617.88235			

R-Square	Coeff Var	Root MSE	Temp Mean
0.816360	7.964202	6.202708	77.88235

Source	DF	Type I SS	Mean Square	F Value	Pr > F
Month	4	7061.127514	1765.281879	45.88	<.0001
Wind	30	1973.863543	65.795451	1.71	0.0362
Month*Wind	53	2082.107962	39.285056	1.02	0.4648

Source	DF	Type III SS	Mean Square	F Value	Pr > F
Month	4	3380.945032	845.236258	21.97	<.0001
Wind	30	2281.515163	76.050505	1.98	0.0112
Month*Wind	53	2082.107962	39.285056	1.02	0.4648

MANOVA by Example

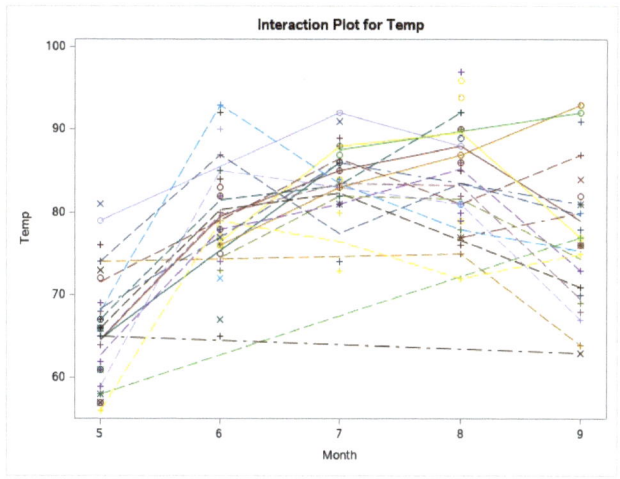

The p-value for the interaction term Month*Wind is 0.46 is not statistically significant (> 0.05).

ABOUT THE AUTHOR

Faye Anderson, MS, PhD is a published author of many peer-reviewed publications and books. She has graduate degrees from Colorado State University and the University of Texas School of Public Health and have been working as a consultant and a statistician for more than twenty years.

OTHER TITLES BY THE AUTHOR

1. Anderson, F. (2019). ANOVA by Example, Hands on approach using R. KDP.
2. Anderson, F. (2016). Categorical Data Modeling by Example, Hands on approach using R, CreateSpace Independent Publishing Platform.
3. Anderson, F. (2016). GeoStatistics by Example, Hands on approach using R, CreateSpace Independent Publishing Platform.
4. Anderson, F. (2016). Hypothesis Testing by Example, Hands on approach using R, CreateSpace Independent Publishing Platform.
5. Anderson, F. (2016). Logistic and Multinomial Regressions by Example, Hands on approach using R, CreateSpace Independent Publishing Platform.
6. Anderson, F. (2016). Statistics by Example, Hands on approach using R and/or Excel, CreateSpace Independent Publishing Platform.
7. Anderson, F. (2016). Survival Analysis by Example, Hands on approach using R, CreateSpace Independent Publishing Platform.
8. Anderson, F. (2017). Biostatistics by Example, Hands on approach using R, CreateSpace Independent Publishing Platform.
9. Anderson, F. (2017). Clinical Trials Statistics by Example: Hands on approach using R, CreateSpace Independent Publishing Platform.
10. Anderson, F. (2017). Time Series Analysis by Example: Hands on approach

using R, CreateSpace Independent Publishing Platform.

www.ingramcontent.com/pod-product-compliance
Lightning Source LLC
Chambersburg PA
CBHW040338220526
45473CB00009B/2724